Socio-Economic Factors Influencing Woodlot Farming in Tanzania

A Case of Mufindi District

Socio-Economic Factors Influencing Woodlot Farming in Tanzania

A Case of Mufindi District

Yona Lumliko
Robert Makorere

ELIVA PRESS

ISBN: 978-99949-8-810-5

Cover Design: Eliva Press
Cover Image: Ingimage
Email: info@elivapress.com
Website: www.elivapress.com

This Eliva Press imprint is published by the registered company Eliva Press Global Ltd. part of Eliva Press S.R.L. Publishing Group

The registered company address is: Pope Hennessy Street Level 2, Hennessy Tower Port Louis, Mauritius
Eliva Press S.R.L. Publishing Group address is: Bulevardul Moscova 21, Chisinau, Moldova, Europe

Socio-Economic Factors Influencing Woodlot Farming Adoption from Crop Farming in Tanzania: A Case of Mufindi District

Yona Lumliko[1] and Robert Makorere[2]

[1] Department of Agricultural Economics and Business, University of Dar es Salaam, Dar es Salaam, Tanzania

[2] Department of Marketing and Entrepreneurship, Mzumbe University, Morogoro, Tanzania

Correspondence: Yona Lumliko, Department of Agricultural Economics and Business, University of Dar es Salaam, Dar es Salaam, Tanzania. E-mail: ylumliko@gmail.com

Abstract

The study stressed to examine socio – economic factors influencing woodlot farming adoption in Mufindi district, Tanzania. The study used questionnaire to collect data from 40 woodlot owners (sample size). The collected data were analyzed using multinomial logistic model. Findings show that the high income generating expectation, land size, and assets acquisitions both influence woodlot farming adoption positively and significantly while education level influence woodlot farming adoption negatively and significantly. Basing on the findings, the study concluded that socio-economic factors that influence woodlot farming positively and significantly are income generation expectation, land size, and assets acquisitions meanwhile education level influence woodlot farming adoption negatively and significantly. Based on the findings, this study recommends that there is a need for government to support the woodlot owners by providing subsidies including inputs such as land, fertilizers, and quality seeds through Mufindi district officials so as to create employment opportunities among majority of rural households and raise their standard of living. If all these recommendations are implemented, then improvement in the income from woodlot farming would ultimately be realized.

Keywords: socio-economic factors, woodlot, woodlot farming adoption, woodlot owners, food security, agroforestry, small-scale forestry, rural livelihood.

1. Introduction

In early 1960s and 1970s, the subsistence farming was a main economic activity by majority of households in Mufindi district for food and cash crops production. Farmers received a great financial, material and technical support from the government in terms of subsidies, agro-machines, fertilizers and quality seeds (Ngailo *et al.,* 2011).

Due to the cut down of government's support to farmers and decline of agricultural market in 1990s for food and cash crops, the majority of households started to think for alternative economic activities such as woodlot farming to generate income which resulted to the massive conversion of agricultural land to woodlot farms (Dewees and Saxen, 1997).

However, the introduction of Structural Adjustment Programs (SAPs) in 1980s and 1990s discouraged the smallholder farmers because it required the privatization of agricultural enterprises which increased the prices of fertilizers and other agricultural inputs, and reduced access to subsidies which made the smallholder farmers to adapt woodlots farming as an alternative economic activity to generate households' income (Reed, 1992; Haule, 2003). Woodlot farming is considered for household income generation because it is cultivated mainly for timber production, paper production, landline telephone poles and electricity transmission poles (Nyadzi *et al.*, 2003) for exchanging.

The majority of households in Mufindi grow different species of trees for different purposes. The most dominant tree species are pines (*Pinus patula*) these are coniferous resinous trees growing 3-80m tall and long lived and typically reach ages of 100-1000 years. They are most useful in timber industry, construction industry, paper miller, saw millers, carpentry, manufacture of plywood and fiberboards. Another tree species grown in the study area is cypress (*Cupressus lusitanica*) which is used in extraction of oil for medicinal benefits such as curing coughs, colds and bronchitis, hemorrhoids and varicose veins. Furthermore, it is used for decoration during Christmas festival in many families across world. Lastly, eucalyptus species (*Eucalyptus myrtaceae*) is mostly used for transmission and telephone poles in the country (Barnes and Wagner, 2004).

However, there are two major landholding woodlot farming systems in practice in the study area which are small-scale and large-scale trees farming. The small scale farming encompasses the farmers with 1-5 acres while above 20 acres are categorized as large scale woodlot farming. The advantages of small-scale is that it is cheap to manage the land in case of diseases, pests and animals control but its major disadvantage is that it may cause low supply of timbers against the demand. (The advantage of woodlot farming in large-scale) is generation of huge income for the households but it is difficult to control fire in case of emergency (FDT Report, 2014).

In Mufindi district, woodlot farming is done in 24,000 hectares (Edwin, 1990). Woodlot farming is growing very high in Mufindi district due to an increase of demand for timbers for building and construction activities in the country. For example, according to Edwin (1990) woodlot farming is done over 24,000 hectares in the district. Another example, the trees harvested are sold to timber traders, saw millers and paper millers which contribute significantly to the households' income (National Bureau of Statistics, 2013).

Despite that the woodlot farming adoption is growing rapidly in the district, currently there are no studies stressed to find out socio- economic factors that influence adaptation of woodlot farming by majority of rural households in Mufindi district yet (Akida and Blomley, 2006). In this view, the reasons caused majority of rural households to adapt woodlot farming as an alternative economic activity to generate households' income (Reed, 1992; Haule, 2003) has not yet been researched. Thus, this study was done to assess socio-economic factors which influence woodlot farming adoption among many rural household farmers especially in Mufindi district, Tanzania.

1.1 Woodlots Situation in the World

Hundreds of millions of people around the globe depend directly on tropical forest ecosystems for both food security and the ecosystem services they provide (Guariguata and Balvanera,2009) A significant proportion of the population of rural people in the world is to a greater or lesser degree forest dependent. Although these people modify forests, they do not, in general, threaten their existence in fact, forest-dependent people have a strong interest in maintaining forests for their survival (Kimaro et al., 2009).

1.2 Woodlots Situation in East Africa

As the rural population keeps on increasing at alarming in East Africa region and the related socio-economic and technological advancement will multiply the demand for and consumption of forest products including timbers for construction projects, fuel wood, saw millers, transmission poles (Dewees and Saxena, 1997).

The demand of forest products can only be met by establishment and management of sustainable forest plantations and woodlots by encouraging the individual people, households, groups, government agencies, public and private companies to engage aggressively in planting trees for commercial purposes (Buyinza and Ntakimanire, 2008). During era the Structural Adjustment Programs (SAPs) in 1970s and 1980s, the East African member countries put emphasis establishment of woodlots as individual or community forestry undertakings, and for environmental protection purposes as a result of ever increasing deforestation and wildfires for farming activities in the region of East Africa (Arnold, 1984).

In traditional practice, wood production in woodlots competes with food production for the same space, undermining the sustainability of production especially on small farms. When farms are large, farmers can allocate parts of their land for growing woodlots. This practice, however, is only feasible if woodlot products are market driven (Dewees and Saxena, 1997).

1.3 Woodlots Situation in Tanzania

The increasing involvement in establishment of private woodlots in Tanzania by individuals has been considered as a long term economic investment with a huge return on capital employed for the majority residents due to increase of demand for timber products especially for construction projects, fuel wood, transmission poles, saw millers, paper millers (Nyadzi, et al., 2003).

The socio-economic development of the district is earmarked by the timber trade. The massive trees planting is done parallel with small scale subsistence farming for food crops especially maize, beans, irish potatoes (FDT, 2014). However, there has been a rapid decrease of arable land to woodland due to the fact that a good number of individuals aggressively concentrate on establishing woodlots despite the fact that it takes more than 10 years for harvesting but people are willing to do that with an expectation of becoming rich after trees harvesting (URT, 2002).

The shift from subsistence crop farming to woodlots farming has been due low prices of cash crops such as coffee, cotton, sugarcane, pyrethrum which in turn has compelled farmers to plant more trees than agricultural crops as an alternative to income generation (URT, 2006). Subsistence farming is very important because it provides food supply to majority people even those who are engaging into woodlots farming need food for survival, they get money from timber products but at the end they have to go to the Supermarket to buy food stuffs for their families therefore despite of huge income that is generated from trees harvesting the subsistence agriculture should not be undermined for the future and sustainable food supply to the households (Ramadhani,et al., 2002).

The population growth rate is high in rural areas this alerts the shortage of food supply in future as the majority are converting the arable land to woodland. The government of United Republic of

Tanzania through the Ministry of Agriculture should raise awareness on the importance of food security so as to avoid the big population to fall in severe famine in future (NBS, 2014).

1.4 Definition of Keywords

As far the research topic is concerned, there are some important key terms that will help grasping the socio – economic factors influencing woodlot farming in Mufindi district, such terms include the woodlot, food security, agro forestry, small scale forestry and rural livelihood.

requirements near irrigation schemes. In coffee-growing areas they are only relevant on waste land that is not fit for other use (Essa, 2004).

1.4.1 Woodlot

A piece of land dedicated to tree planting usually located around a household or within a village. A woodlot may be owned by an individual, a household or by a community. A household woodlot is therefore a piece of land where the household members plant trees and have the rights over the trees planted and make their own management decisions (Chikoko, 2002).

Woodlots are particularly relevant in areas where light-demanding crops are grown, e.g. in areas where maize or sugarcane are dominant crops. Woodlots are also relevant for meeting wood requirements near irrigation schemes. In coffee-growing areas they are only relevant on waste land that is not fit for other use (Essa, 2004).

1.4.2 Food Security

Food security exists when all people, at all times, have physical and economic access to sufficient safe and nutritious food that meets their dietary needs and food preferences for an active and healthy life (WFS, 1996). Also it refers to availability at all times of adequate world food supplies of basic foodstuffs to sustain a steady expansion of food consumption and to offset fluctuations in production and prices" (UN, 1975).

By accounting for vulnerability, food security policies and programs broaden their efforts from addressing current constraints to food consumption, to include actions that also address future threats to food security (Dilley and Boudreau, 2001).

1.4.3 Agroforestry

Agro forestry is a combination of agriculture and forestry, resulting in better management of natural resource and sustainable use of land. In other words, agro forestry is a method of farming that allows trees and shrubs to grow along with crops and/or livestock, therefore blending agriculture and forestry in the same production system. Agro forestry is increasing being recognition of a reasonable option of overcoming the global challenges of climate change mitigation, adaptation, food security and household income (Nair et al., 2009).

The farmers benefit in many ways with a well-designed, well planned and well executed agro forestry system. The presence of trees and shrubs aid the crops in making better utilization of the soil nutrients and light, hence resulting in better production as compared to mono-cropping method (Garrity and Stapleton, 2011).

Agro forestry systems provide an option of earning extra income to the farmers. Diversification between forest and agricultural crops reduces labor and resource costs along with the risks undertaken by farmers. Mixing long term forest crops with yearly agricultural income makes the overall system more profitable. Lastly forest provides food and shelter for the livestock which is beneficial in many ways (Okia et al., 2009).

1.4.4 Small – Scale Forestry

Small-scale forestry – here meaning all activities related to forest ownership, management and enterprise can contribute significantly to social and economic development when the circumstances are right (Ngaga, 2011). The nature of small – scale forestry differs from one place to another, what is considered as small in one place might considered as large, but small – scale forestry consists of single or smaller number of planting blocks, non – professional management and often a lack of silvicultural skills with poor planning for future forest products marketing (Harrison et al, 2000).

1.4.5 Rural Livelihood

A livelihood comprises the capabilities, assets (stores, resources, claims and access) and activities required for a means of living. The term 'livelihoods' has become increasingly important in development theory and practice, as it is seen to encompass a wide range of concerns, and to allow inclusion of the broad range of people's activities and assets in considering how they support themselves, rather than focusing more narrowly on economic, income-generating or formal activities (Chambers and Conway, 1992).

2. Conceptual Framework

The conceptual framework is governed by the theory of transaction costs. Transaction cost economics, unlike traditional neoclassical economic theory, recognizes that commercial activities do not occur in frictionless economic environment (Williamson, 2000). Instead, it posits that there is costs top carrying out any exchange. However, the theory insists that exchanging costs must be reduced for traders to earn lucrative profit.

In this view, the study applied the Transaction Cost Theory to see if the study would support or divert the theory. According to Institutional Analysis and Development (IAD) framework, socio-economic factor is one of the institutional factors associated with marketing or exchanging costs. According to Kirsten *et al.,* (2012), economic actors who have unfavorable socio-economic factors in term of business, also chances of such actors to incur high transaction costs is high and consequently may fetch low income.

Based on the aforementioned factors, this study illustrates that there is existing relationship between woodlot farming adoption as a source of income growth and socio-economic factors of woodlot owners as shown in conceptual framework in Figure-1.

The conceptual framework further shows that the woodlot farming adoption depends on socio-economic factors and other factors as can be seen in the Figure-1 below:

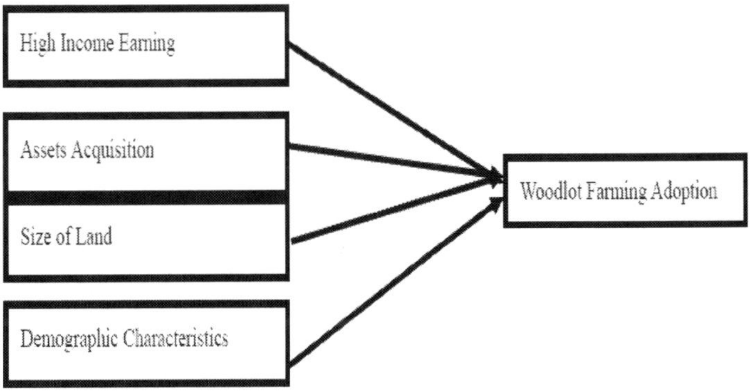

Figure 1. Conceptual framework

According to conceptual framework (Figure-1) socio-economic factors is one of the intrinsic factors that influence woodlot farming adoption (Kirsten *et. al.,* 2012; Mnyawi and Benedict, 2014). Based on the socio-economic factors, woodlot farming adoption is directly related with gender, age, education level, amount of wealth. In another words, this is to say that, figure-1 shows that socio – economic factors such as demographic characteristics (gender, age, education level), expectation of high income earning (which is measured in amount of money in Tanzanian Shillings earned by woodlot farmers in a year), assets acquisitions (e.g. types of assets acquired such as motor vehicles, houses, saw mills, chain saws and generators) and size of land (which is measured in acres of land with trees owned by woodlot farmers) directly relate with woodlot farming adoption. According to literature review, this means that high income earning expectation, assets acquisition, size of land, and demographic characteristics of household heads both have influence on woodlot farming adaptation (Singunda, 2010; FDT, 2014). Despite all these studies, however, still little is known on the socio-economic factors influencing woodlot farming adoption especially in Tanzania. In this view, it is therefore important to understand socio-economic factors influencing woodlot farming adoption using Transaction Cost Theory and Institutional Analysis and Development(IAD) framework approach.

3. Research Methodology

3.1 Study Area

This study was conducted in Mufindi district particularly in three divisions Ifwagi, Kibengu, and Kasanga which comprises of the villages such as Nyololo, Igeleke, Igowelo,Ihefu, and Isalavanu where the data were collected. This district was selected because frequency of farmers to transform farm uses from crop farming to woodlot farming is high in Mufindi district. As noted by Forest Development Trust Survey Report (2014) that, there is high rate of shifting from crop farming to woodlot farming in Southern Highlands of Tanzania including Mufindi district in Iringa region. In Mufindi district, 37% of tree growers transformed crop farms to woodlot farms while 13% of tree growers converted their woodlot farms to crops farms (Singunda, 2010; Chikoko, 2002). This led the researcher to study socio – economic factors that influence woodlot farming adaptation in Mufindi district.

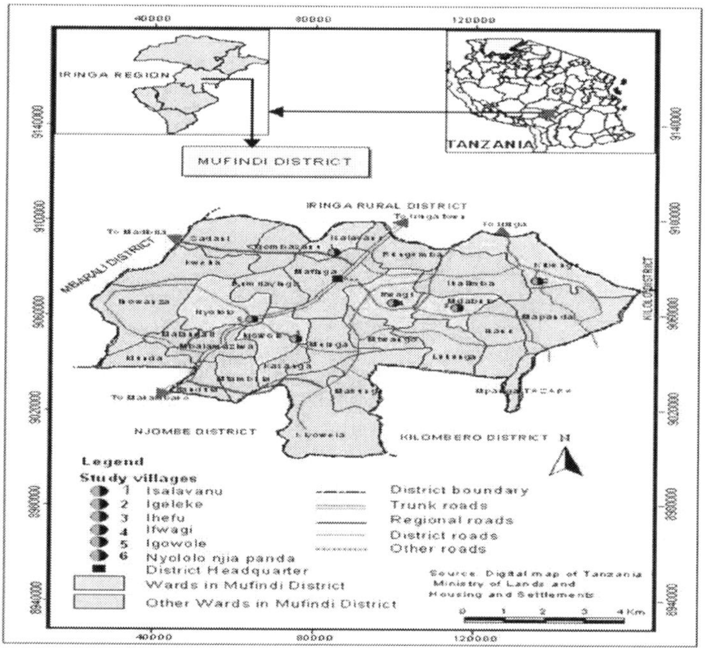

Figure 2. The map of Mufindi district showing the villages where the study was conducted

3.2 Sampling and Data Collection Method

The target population of the study involved woodlot farm owners in Mufindi district in Iringa region. The study selected woodlot farm owners for providing firsthand information on the matter understudy including reasons for transform crop farming to woodlot farming. In this study, simple random sampling procedure was used to get representative sample of 110 woodlot farmers for this study. The sampling procedure used is that, the selection procedure was as follows; woodlot farmers were visited at their household or working places and were informed the purpose of the study so as to get their consent of participating in the study before being selected. After getting their consent, woodlot farmers were requested to participate in this study. However, the selection was basing on the list of woodlot farmers obtained from Village Executive Officers (VEOs). By using rotary approach, woodlot farmers were selected from the list for this study.

Through this selection procedure, the researcher managed to select a sample size of 40 woodlot owners for inclusion in this study. However, the criterion the researcher used to arrive at this number is borrowed from Bailey (1994), that a sample or sub-sample of 30 respondents is bare minimum for studies in which statistical data analysis is to be done for a study like this regardless of population size, hence the study fulfilled Bailey's (1994) requirement.

The data collection process comes after the sampling stage in the research process (Churchill and Lacobucci, 2002). In this particular study, a survey method involving administered questionnaires was used for data collection. Self-administered questionnaire as an approach of data collection was used for guaranteeing a high response rate from the respondents (woodlot farmers).

One set of a questionnaire was prepared and pre-tested as the main data collection instrument to woodlot farmers. Technically, the researcher asked the consent of the sampled woodlot farmers to participate in the study and afterwards the sampled woodlot farmers (respondents) were informed about the objectives of the study and were then required to confirm their willingness and availability to participate in the study. Eligible woodlot farmers included in this particular study are those woodlot farmers who are heads of household.

The study measured validity of data collected to testify authenticity of the data collected to verify that the findings are safe for consumption. Pilot study was carried out before field survey to measure data collection instruments. During pilot study, the study tested the ability of the instrument to measure the constructs as validity measurement. Based on the outcome of the pilot study, some questions were restructured while others were dropped out. This procedure was done to the questions which were seen to be vague and not practical in the questionnaire. Moreover, observation method was used in this study. Materials used for house construction (wall, roof and floor), the size of land with woodlots, and other available assets (motor vehicles, motor cycles, generator, and milling machines) were observed during the survey.

Nevertheless, reliability was not tested during pilot study, but was taken care during data collection phase since validity (Note 1) is a precondition of reliability (Gujarati, 2006; Pindyck and Rubinfeld, 2010; Creswell, 2011); Tatham *et al.,* 2007). Moreover, everything that is valid is reliable, but not everything that is reliable is valid (Creswell, 2011; Tatham *et al.,* 2006).

During the field survey, the study took some measures to control validity during data collection phase as stated earlier. *First,* "training" was conducted to the research assistant before starting the field survey. The training was intended to make the enumerator more familiar with the constructs used in the questions in the questionnaires. This helped the researcher to address the problem of validity in the study. *Second* every day after field survey, the researcher and the research assistant "cross-checked" the filled questionnaires to see if the constructs were clearly understood and measured. *Three,* the study used "administered questionnaire procedure to control misinterpretation of questions that might have occurred among the woodlot farmers (respondents).

3.4 Analytical Technique

This particular study applied logistic regression method to analyze the socio-economic factors that influence woodlot farming adoption from crop farming in Tanzania particularly Mufindi district. The selection of this method in this study was influenced by the fact that logistic regression proved success in assessing socio-economic factors influencing woodlot farming adoption in many other previous studies. The variables for inclusion in this study were also selected from the past experience of studies based on their relevance in influencing woodlot farming adoption.

Therefore, logistic model was used to assess socio-economic factors that influence woodlot farming adoption among crop farmers in Tanzania. The model was employed to predict the power of predictors in explaining the dependent variable (woodlot farming adoption). The study assumed that crop farmers may transform part or all of his/her land to woodlot farming or not. However, before carrying out regression analysis, the study assessed collinearity of independent variables before conducting regression analyses.

Therefore, the specification of respective logistic regression model is presented below:

$$Logit(Y_i) = Exp(\beta_0 + \beta_1 X_1 + \beta_2 X_2 + \beta_3 X_3 + \beta_4 X_4 + \beta_5 X_5 + \beta_6 X_6 + e)$$

Whereby;

Y_i=Woodlot Farming Adoption (Dependent variable)

X_1= Gender of a farmer

X_2= Age of a farmer

X_3=Education level of a farmer,

X_4 = Annual income earned by a farmer from woodlot farming

X_5 = Land size own by a farmer

X_6 = Assets acquisition by a farmer

e = Standard Error

β_0 = Intercept

$\beta_1 .... \beta_6$ = Coefficient of determinant

Based on the aforementioned logistic regression model above, the study generally assumed that woodlot farming adoption is influenced by certain intrinsic factors which include gender of a farmer, age of a farmer, education level of a farmer, annual income of a farmer, size of land owned by a farmer, and assets acquisition by a farmer.

(a) **Measurement of dependent variable:** Dependent variable (DV) is woodlot farming adoption in the study. In this study, the dependent variable (woodlot farming adoption) was measured by asking the respondents whether they adopted woodlot farming from crop cultivation or not. The unit of measurement used was "Yes" if Adapt, "No" if Not Adapt.

(b) **Measurement of the Independent Variables:** In this study, six independent variables were measured. Table-1 presents the summary of independent variables which influence woodlot farming adaptation among majority of rural households in Mufindi district.

Table 1. Summary of Independent Variables Influencing Woodlot Farming Adaptation in Mufindi district (n=110)

Variable Symbol	Variable Description	Nature of Variable	Unit of Measurement	Expected Sign
X_1	Gender of a farmer	Binary	Dummy: (0 for Male and 1 for Female)	–
X_2	Age of a farmer	Continuous	Years of woodlot farmer	+
X_3	Education level of a farmer	Continuous	Years of respondent in school	+
X_4	Average income of a farmer	Continuous	Total income per year from woodlot farming	+
X_5	Land size own by a farmer	Continuous	Hectares of woodlot farm	+
X_6	Assets acquisition by a farmer	Continuous	Type(s) of asset(s) owned	+

3.5 Technical Considerations

At the outset, this study assessed technical issues related to multicollinearity and regression results validity. Meeting the assumptions of regression analysis is essential in ensuring that the results obtained are (Katani, 1994) truly representative for the researcher to be able to generalize the results.

Therefore, the following examination of basic assumption of regression analysis was done, namely multicollinearity; and further testing of the overall relationship after model estimation.

4. Findings and Discussion

The paper aimed to assess the socio-economic factors influencing shifting from crop farming to woodlot farming in Mufindi district. Empirical evidences have been showing that majority of rural households own by crop farmers experience income poverty due to unfavorable socio-economic factors surrounding them compared with those farmers who had woodlot farms. As noted by Bailey (1994), unfavorable socio-economic factors are associated with high transaction costs, which would reduce income from exchanging. Therefore, socio-economic characteristics of woodlot farmers were reviewed prior the study applied logistic regression method to explain socio-economic factors that influence woodlot farming adoption as can be seen in the next sub-section below:

4.1 Socio-economic Characteristics of Respondents

The paper sees it is relevant to examine socio-economic characteristics of woodlot farmers to determine features of the respondents included in the study. Empirical evidences have been showing that socio-economic factor is one of the major factors influence income growth of any economic actors (Mnyawi and Benedict, 2014; Kirsten *et. al.*, 2008; Hobbs and Young, 2001). In this view, the following are the intrinsic socio-economic characteristics assed in the paper:

Table 2. Socio-economic characteristics of respondents (n=110)

Characteristics	**Category**	**Frequency**	**Percent**
Gender of respondents	Male	59	54
	Female	51	46
Age of respondents	Less than 40 Years (Youth Age)	18	16
	41-54 Years (Adult Age)	25	23
	55+ Years (Old Age)	67	61
Education level	Never gone to school	19	17
	Primary level	41	37
	Secondary level	26	24
	Non-degree level	11	10
	Degree level	13	12
Woodlots size	Less than 9 Acres	74	68
	10 – 19 Acres	21	19
	20 – 29 Acres	8	7
	30+ Acres	7	6

Tree species	Pines	64	58
	Eucalyptus	35	32
	Cypress	11	10
Average Income	Less than Tshs. 5 Million	59	54
	Tshs. 6 - 10Million	35	32
	Tshs. 11 - 20 Million	8	7
	Tshs. 21 and above Million	8	7

4.1.1 Gender of Respondents

Table-2 shows that the study was not biased because gender balance i.e. male and female was seriously considered. For instance, the Table-2 shows that male were 59(54%) and female were 51(46%) participated in this study. However, empirical evidence shows that most of the woodlot farm owners were male; if you based to the African tradition, normally female are staying home to take care of the family, fetch firewood and water while head of the house (male) is fetching bread for the family.

4.1.2 Age of Respondents

In Mufindi district, it was observed that both youths and elders are actively engaging in woodlot farming as main business activities in their villages. In this study, the age of respondents was categorized into the following groups as can be seen in Table-2 above: less than 40 years, 41 to 54 years, and 55 years and above. According to Table-2, the highest age group is 55+ years which is equivalent to 67 (61%) and the lowest age group less than 40 years which is equivalent to 18 (16%). The implication here is that majority of people who converted crop farms to woodlot farms are older people due to their long time experience in crop farming and woodlot farming than any other age group in the study area.

Considering that empirical evidence shows that age has a significant relationship with the growth of income in business (Makauki, 1999). Theoretically it is believed that person who has higher age is someone with high business experience and knowledge, and consequently can generate more money using his/her business experience and knowledge.

4.1.3 Education Level

The findings in Table-2 shows that 41(37%) respondents had primary education while 26(24%) had secondary education. Based on these findings, majority of respondents (woodlot farm owners) had primary education. Theoretically, this implies that these woodlot owners have limited entrepreneurial skills and business education. Consequently, the level of illiteracy is very high. Therefore, level of influencing big deals is very minimal and hence getting low business income. Literature shows that low level of education might have affected growth of income of woodlot owners in the study area.

4.1.4 Woodlot Size

Based on the findings, the study shows that the land size owned by the respondents differs in term of hectares as Table-2 shows. The findings show that 74(68%) respondents interviewed were owning land less than 9 acres, 21 (19%) respondents interviewed were owning land between 10 – 19 acres, 8 (7%) respondents interviewed were owning land between 20-29 acres, and 7(6%) respondents interviewed were owning land more than 30 acres. This implies that majority of the rural people in Mufindi district have started to convert their crop farms to woodlot farms as 74(68%) respondents interviewed have converted their crop farms to woodlot farms.

4.1.5 Average Income per year from Woodlot farming

Based on the findings, one of the reasons why majority of respondents in Mufindi district adapted woodlot farming is to improve their income. As Table-2 shows that, 59(54%) respondents earned less than TZS 5 Million, 35(32%) respondents earned between TZS 6 – 10 Million, 8(7%) respondents earned between TZS 11 – 20 Million, and 8(7) respondents earned TZS 21 Million and above. The findings imply that majority of woodlot farmers earn less than Tshs. 5 Million per annum.

4.1.6 Tree Species

Based on the findings in this study, the main tree species grown in Mufindi district are pines, eucalyptus and cypress.

The results in Table 3 below shows that 58.2% of the respondents grew pines, 31.8% of the respondents grew eucalyptus while only 10% of the respondents grew cypress. It can be concluded that pines are in high demand than other species due to construction needs and paper millers. It was also reported by FAO (2001) that eucalyptus may deplete the soil nutrients in the area grown. Furthermore, pines are often grouped together in a landscape and they are highly demanded for construction projects across the country. In Tanzania, most of rural people often prefer to decorate their premises using cypress branches during Christmas day. The following Table 4.5 presents the responses on the tree species grown in Mufindi district.

Table 3. Dominant Trees Species in the Study Area (n=110)

Tree Species	Frequency	Percentage
Pines	64	58.2
Eucalyptus	35	31.8
Cypress	11	10

Figure 3 Pines woodlot

The Figure 3 above shows the pine trees which are resistant against diseases, decay and rot hence they are preferred for roofing by majority of construction industries in the country.

Figure 4 Eucalyptus woodlot

The Figure 4 above shows the eucalyptus trees which are mainly used for transmission poles as grown straight and tall than pines and cypress, brewing boiled eucalyptus leaves can help to prevent or cure diabetes. Also eucalyptus trees are planted once in the farms after harvesting the remaining roots can grow again hence no need of seedlings for the scheduled planting.

Figure 5 Cypress woodlot

The figure 5 shows the Cypress trees are quickest growing species though are not very much resistant against diseases but are mostly used by majority of rural households in Christmas as the result this species is now known locally as Christmas tree. Furthermore, its cones and needles can offer medicinal treatment against arthritis and gout as reported by respondents.

4.1.7 Trees Planting

The results in Table 4 indicates that 76.4 % of the key informants confirmed that the trend of trees planting is increasing rapidly with the population increase which results into acute shortage of land,while 23.6% of the key informants confirmed that the trees planting rate is decreasing.

Table 4 Trend of Trees Planting

Trend	Frequency	Percentage
Increasing	84	76.4
Decreasing	26	23.6

4.1.8 Trees Harvesting

In Table 5 below it was note that 33.3%confirmed that the trend of trees harvesting is excessive while 66.7% confirmed that trend of trees harvesting is normal. However, many woodlot owners have not accessed education on proper harvesting methods which eventually resulted into decrease of annual yields due to poor harvesting methods used in the study area. Therefore, the government should provide proper harvesting education to woodlot farmers through local government authorities (LGAs). Table 4.7 below presents the trend of trees harvesting in the study area.

Table 5 Trend of Trees Harvesting

Trend	Frequency	Percentage
Normal	73	66.7
Excessive	37	33.3

Figure 6 Trees harvesting

The Figure 6 above shows how trees are harvested unprofessionally, lack of education on trees harvesting is a challenge. Many trees which are useful left behind during harvesting as the result of poor equipment and methods used. Axes were reported by respondents mainly to be used for falling trees down instead of modern machines such as chain saws.

4.1.9 Price Trends of Timbers, Logs and Transmission Poles

The data in Table 6 below revealed that 77.3 % of respondents reported that the price is increasing while only 22.7% of the key respondents confirmed that the price is decreasing. This implies that the price of timbers, logs and transmission poles is increasing time after time.

Table 6 Price Trends (n=110)

Trend	Frequency	Percentage
Increasing	85	77.3
Decreasing	25	22.7

4.4 Analysis on the Major Reasons for Woodlot Farming Adaptation in the Study Area

Woodlot farming plays an important role in majority of rural households in Mufindi district. The increased size of household members also implied the increased struggles to secure new sources of income for subsistence needs. However, woodlot farming is currently seen as an alternative survival strategy for majority of rural household in study area. This is because of poor and declining market of cultivated crops over past 5 years (Singunda, 2010). The following Table 4.9 shows the reasons as to why majority of rural households adapted woodlot farming as an alternative socio-economic activity for satisfaction of subsistence and secondary needs.

4.2 Regression Analysis Results

In this study, logistic regression model specified socio-economic factors that influence woodlot farming adoption. The regression output is shown in Table-.7 The findings indicate that three out of six independent variables [i.e., petty traders' education level (PTEDUC), petty traders' business experience (PTBEXP), and petty traders' business assets (PTBASS)] had significant effect on the household petty traders' income growth as shown in Table-7.

Table 7. Results of logistic Regression Analysis on the Socio-Economic Factors Influence woodlot farming adoption (*n* =110)

Variable	β	SE	Sig.	Exp (β)
Gender of respondents(X_1)	-1.734	1.344	0.197	0.177
Age of respondent(X_2)	0.31	0.752	0.967	1.031
Education level of the respondents (X_3)	-1.170	0.634	0.045**	0.311
High income expected by respondents (X_4)	3.313	1.334	0.014**	0.713
Land size owned by respondents(X_5)	0.500	0.465	0.023**	1.051
Assets Acquisition by respondents(X_6)	0.017	0.176	0.019**	1.017
Constant	1.440	1.868	0.070**	1.220
Cox & Snell R^2= 0.754				

Note: ** Significant at 5%

Nagelkerke R^2= 0.905

-2Log Likelihood (-2LL) = 24.846

4.2.1 Gender of the Respondents

Survey results in Table-3 indicate that coefficient of gender was insignificant but positively influenced woodlot farming adoption decision; this result is in contrast with the prior expectation. But however, this result is in line with the result shown in the study done by Foy and Pitcher (1999) and Katani (1994) that, gender has negative sign and is insignificant as woodlot farming does not increase with more participation of women. Since the empowerment of women seems to be fundamental in achieving rural household income poverty reduction.

On the other hands, gender is important as it reflects the traditional biasness in relation to land ownership and rights. Normally household has right to inherit land or trees; in most cases women have no right to inherit land or trees. As a result, gender ends up having a negative effect on performance of woodlot farming because women are the main users of waste forestry products such as charcoals and firewood.

4.2.2 Age of the Respondents

Based on the survey results, the findings of this study in Table-3 shows that the coefficient of age of the woodlot owner was insignificant but positively influenced woodlot farming adoption decision; this result concur with the prior expectation. This implies that woodlot farming adoption is positively influenced by age of the head of household. The notion here is that the older the age of householder' head is, the less productive s/he gets and hence the lower the income generated thus needs to convert crop farm to woodlot farm for more income generation. This result agrees with theoretical view that older have more personal contacts, allowing trading opportunities to be discovered at low cost, and are likely to experience lower transaction costs and use more marketing channels (Lyne, 2002), and get higher income from woodlot farming.

4.2.3 Education Level

Based on the findings, education is one of the strongest determinants of woodlot farming adaptation in Mufindi district. This is because respondents with higher education were found to have more annual income earning from woodlot farming compared to respondents who have below higher education level. This implies that woodlot farmers with at least higher education level have great chance to perform well in woodlot farming and earned more income.

Education plays an interesting role as a determinant of activity income. The coefficient of education level of respondents was significant and negatively influence woodlot farming adoption decision as noted in Table-3. This result does not conform to a prior expectation that the higher the level of education of the household head is, the higher the household income in the woodlot trade exchange activities. This finding agreed with the finding in a study done by Ugwumba *et al.,* (2010) who observed that household heads who have higher education can combine many viable enterprises and can be more efficient in woodlot farming adoption and consequently realize more income.

In addition, the low level of education of household heads, coupled with their inability to communicate in the business language (*e.g.* lack of bargaining skills or inability to identify poor contract), also contribute to high transaction costs among household heads who engage into woodlot farming. High transaction costs in marketing of commodities often exclude poor woodlot farmers from participating in growth opportunities (Lyne, 2002).

4.2.4 High Income Expected

Based on the survey results, the findings of this study in Table-3 show that the coefficient of income growth aspiration was significant and positively affected by woodlot farming adoption decision; this result concur with the prior expectation. This implies that an increase of woodlot farm size goes together with an increase of woodlot owners' income per year.

4.2.5 Land Size Owned

Based on the findings, the coefficient of land size owned by respondents was significant and positively influenced woodlot farming adoption decision as noted in Table-3. This result conforms with a prior expectation that the higher the amount of land size owned by farmers, the higher the chances of converting crop farms to woodlot farms. This is to say, the bigger the size of land owned, the more the space would be available for woodlot farming adaptation, and tree planting by the household. Here, the study assumed that, size of land determines the extent to which the majority of rural household in Mufindi district would adapt the woodlot farming.

4.2.6 Assets Acquisition

Based on the findings, the coefficient of assets acquisition owned by respondents was significant and positively influenced woodlot farming adoption decision as noted in Table-3. This result conforms also with a prior expectation that the higher the amount of assets acquisition expectations owned by respondents is, the higher the convention rate of crop farms to woodlot farms.

The study however revealed that machineries and equipment (assets) had influenced respondents to convent crop farms to woodlot farms. For instance, respondents who owned saw mills and paper mills invested in woodlots so as to obtain logs and timbers and exchange at high value.

4.3 State of Housing Materials of Woodlot Owners

The information on the state of housing materials used for construction is useful since it reflects the household socio economic status. The results from the study indicated that 65.5% of the respondents' houses walls were built by bricks, 28.2% made of blocks, 4.5 % made of timbers and 1.8 % made of mud/soil. Furthermore, 71.8% of the respondents' house roofs were made of iron sheets, 21.8% made of timbers, 1.8% made of thatches/grass and 4.5% made of tiles. Nevertheless, 10% of the respondents' house floors were made of ceramic tiles, 9.1% made of mud/soil and 80.9% made of cement. This implies that woodlot farming adaptation has improved the rural livelihood of majority household as many houses' walls are made of bricks, roofs made of iron sheets and floors made of cement, this is a social economic progress as far as the rural life is concerned. The following Table 8 presents the data collected on the housing materials of the woodlots owners.

Table 8 Housing Materials (n=110)

Variable	**Material used**	**Frequency**	**Percentage %**
Wall	Brick	72	65.5

	Block	31	28.2
	Timber	5	4.5
	Mud/Soil	2	1.8
Roof	Iron sheets	79	71.8
	Timber	24	21.8
	Thatches/Grass	2	1.8
	Tiles	5	4.5
Floor	Ceramic tiles	11	10
	Mud/Soil	10	91.
	Cement	89	80.9

4.4 Another Reasons Influencing Woodlot Farming Adoption

Woodlot farming plays an important role in majority of rural households in Mufindi district. This is because of poor and declining market of cultivated crops over the past 5 years (Singunda, 2010). However, there are reasons which entail why majority of rural households adapted woodlot farming and consider them as an alternative socio-economic activity for satisfying their subsistence and secondary needs as shown in Table 9.

Table 9. Other reasons that influence woodlot farming in the district (n=110)

Reasons lead to woodlot adoption decision	**Frequency**	**Percent**
Poverty reduction	33	30
Supply of timbers, logs, and transmission poles	19	17
Environment protection	5	5
Self-employment	10	9
Government policy	8	7

4.4.1 Poverty Reduction

Based on the findings in Table-4 above, woodlots provided a wide range of resources and benefits to poor people, from providing land for agricultural conversion, to timber, agro forestry and non-timber forest products and ecosystem services. The woodlot owners draw substantial parts of their subsistence needs from forest produces. Furthermore, woodlot farming generates the so-called forest environmental income which refers to income (cash and non-cash income) derived from extraction from non-cultivated forest sources. The non-cash income includes food, fodder, energy, house-building materials, and medicine. Based on the findings, 30% of respondents interviewed adapted woodlot farming because of poverty reduction. This will be achieved by selling of logs, timbers and transmission poles.

4.4.2 Supply of Timbers, Logs and Transmission Poles

Findings depicted that 17.3% of respondents adapted woodlot farming for the purpose of supplying timbers for construction industry, logs for saw millers and paper millers as well as transmission pole for national grid electrical service lines as shown in Table 4.9. The eucalyptus is the best used tree species for transmission poles. A study by Singunda (2010) showed than 35% of family tree growers in Mufindi district are new (less than 8 years of engagement in tree planting), most of whom have been attracted by the growing demand for wood raw materials.

Woodlot farming is reportedly ranked as the first economic activity by most of the participant farmers in Mufindi district. However, during the recent past, income generated from selling tree products has shown a considerable increase thereby motivating other farmers to join woodlot farming as a financially attractive activity. Woodlot farming activities in most villages in Tanzania is carried out on the individual farmers' initiatives and species grown for commercial purposes include pines, eucalyptus and cypress.

4.4.3 Environment Protection

The findings in Table 4.9 above, findings showed that 4.5% of respondents adapted woodlot farming for environment protection. This is the smallest percentage compared to other reasons. This implies many woodlot farmers in Mufindi lack awareness on the contribution of woodlots to environment protection education. Furthermore, the diversity of goods and services provided by woodlots serve as a significant safety net source or a gap-filler for the rural poor in times of instability, scarcity. The research in developing nations shows that forests provide "natural insurance" in the form of alternative sources of income and subsistence to help cope with shocks (Kajembe and Luoga, 1996). In particular, forest income diversifies the income portfolio for all groups, but is particularly important for poor, rural households. The woodlots also provide safety nets in the sense that they can reduce environmental impacts of events such as floods and landslides (i.e. regulating ecosystem services).

4.4.4 Self Employment

As noted in the study, the woodlot farming served as a self-employment for majority of rural household. Due to the problems in labor market and illiteracy of majority of rural household in Mufindi district, it is quite very difficult for them to engage in professional jobs but rather the woodlot farming adaptation becomes significant as self-employment for them. The Table 4.9 indicated that 9.1 % of respondents adapted woodlot farming for self-employment. The woodlot owners used household labor for planting and weeding the trees in their woodlot farms.

4.4.5 Government Initiatives under the National Forest Initiative Schemes

The results in Table 4.9 above show that 7.3% of the respondents embarked woodlot farming as the result of national forest initiatives pressure insisting on the importance of trees planting. Also, the government has declared a national day for tree planting (the First of April each year) which is earmarked by each region, district, institution and organization.

It is not clear how many trees have been planted through this campaign and what the survival is, but the contribution is significant. Tanzania's government first launched tree planting campaign in the 1970s and later in the 2000s to ensure every district council plant at least 1.5 million trees annually. Also the government established Tanzania National Forest Policy 1998 and Tanzania National Forest Act 2002 to enforce the importance of trees planting in the country for environment protection and adaptation and mitigation of climate change.

4.5 *Challenges encountered in woodlot farming*

Woodlot farming like other agroforestry projects in Mufindi district encounteres various setbacks which eventually lead to the low yield of logs, or poor quality of timbers from the logs due to the fact that the trees that have been hit by diseases do not grow well as expected by the woodlot farmers.The major and threatening setback was found to be wildfires which was noted to be 40.9% once this happens no single tree is left safe the woodlots without fire breaks are totally consumed and damaged which increase the loss in the investment.

Table 4.14 Setbacks encountered by woodlot farmers in the district (n=110)

Challenges	Frequency	Percentage
Wildfires	45	40.9
Lack of enough land	15	13.6
Lack of proper education on trees planting	13	11.8
Population increase	7	6.4
Bad weather conditions	4	3.6
Trees diseases	11	10
Poor road infrastructure	6	5.5
Trees theft	9	8.2

4.5.1 Wildfires

Table 4.14 above shows that frequent wildfire is among the challenges that face woodlot farming activities in the study area. As forest fires reduce plant biomass and litter, thereby altering the energy, nutrient and water fluxes between the soil, plants and atmosphere. These changes in turn may affect the long term nutrient status and productivity of the system and consequently population structure of a species, the composition of communities and ultimately, the probability and characteristics of future fires (Ngaga, 2011). The ranges that have burned are habitats for different species with restricted home range. The effect of fire on the vegetation is mostly related to its impact on the soil, water, atmosphere and their interconnectedness. Therefore, to control the effects of wildfires, the woodlot owners are advised to use firebreaks along their farms as well as avoiding improper methods of honey harvesting such as the use of fires (Zolho, 2005).

4.5.2 Lack of Enough Land

Due to the population increase in Mufindi district as reported by key informants, the competition for land is increasing as it has been driven mainly by demand for woodlot farming expansion. However, the competition for land between forest plantations and agriculture is intense. It is an established fact that land is the world's most valuable resource and public scrutiny of land is becoming more intense with each passing year. With increasing population, the competition on land for agriculture and development is ever increasing (Nyadzi et al, 2003).

For a woodlot farming investment to be commercially viable a large area is required. The size of the land required will vary with the objective of the woodlots. It would always be desirable to have a single contiguous piece of land area and if not, the required land area should just be in about two or three nearby parcels only. This is to ensure easy and efficient management of the activities. It is always preferable that the acquired land should near to basic amenities and relatively accessible mainland transportation infrastructures (roads and railways) within an economic range to a processing mill or market (Muyanga and Jayne, 2014).

4.5.3 Lack of Proper Education on Trees Planting

Most farmers do not have prior technical knowledge on tree farming other than the informal knowledge either gained through experience or from friends and neighbours. Also, they lack the necessary knowledge on the need of proper land preparation prior to field planting, and therefore, most of the farmers do not prepare the land adequately before planting. The lack of knowledge is also noted in planting spacing adopted by the farmers, and not following the various management schedules. This is partly due to inadequate technical support and extension services, especially in timber planting and harvesting activities (Kajembe and Luoga, 1996).

4.5.4 Population Increase

Apart from respondents' response. It was also reported by key informants (DFO and VEOs) that population increase was among the challenges that faced the woodlot owners in Mufindi district due to the fact that the household head needs more children for working in the tree farms. The population increase may also lead to food insecurity in the district as majority of rural households practicing woodlot farming and few of them cultivated crops. As agriculturally based population density increases in and near forested areas, the strongest relationship between population growth and deforestation occurs, as local people and young migrant families arrive at the forest frontier and clear land to provide more area for subsistence farming (Cochet, 2004).

4.5.5 Bad Weather Conditions

The bad weather condition was reported as another challenge that faced woodlot owners in Mufindi. Climate change is impacting the world's ecosystems and it is expected that the magnitude of these impacts will increase along with temperatures over this century. Many trees species and ecosystems may not be able to adapt as the effects of global warming and associated disturbances drought, strong winds and insect outbreaks, are compounded by other stresses such as land use change, overexploitation of resources, pollution and fragmentation of natural systems. The forest productivity and species diversity typically increase with increasing temperature, precipitation and nutrient availability, although species may differ in terms of their tolerance (Deressa, et al., 2009).

Figure 7 Pine trees hit by strong winds

Figure 7 above shows the main problem reported to challenge woodlot farming in the area of study which was strong winds that destroyed the trees and them fall down and dry before the time of harvest. One falling tree can destroy some other trees that are nearby hence reducing the annual yields.

4.5.6 Poor Road Infrastructure

Poor road infrastructure hindered the transportation of trees, timbers, logs, transmission poles from the farms to the saw mill and paper mills. During harvesting the woodlot owners experienced difficulty in transporting forest goods as many roads are not passable by long vehicles with cargo. The district, urban and feeder roads are the responsibility of local government authorities (LGAs) who are under the oversight of the Prime Minister's Office Regional Administration and Local Government (PO-RALG). PORALG has an oversight function of preparing policies and strategies in consultation with stakeholders as well as coordinating, monitoring, and providing support to LGAs in road works implementation activities (URT, 2006).

Figure 8 Road heading to woodlot farms in Mufindi

Figure 8 above shows that many roads in the study area are muddy as observed by the researcher but also reported by respondents. These roads which are not passable during rainfall season hence harvesting during rainfall was not possible due to the fact the logs can be transported from woodlot farms to the saw mills and paper mills.

4.5.7 Trees Diseases

The woodlot owners in Mufindi district reported trees diseases as a challenging faced woodlot farming. The persistent diseases are fungi and insects that destroy trees which are in growing stage. However, the quality of woodlots in Tanzania means that the country is extremely susceptible to diseases and pests arriving from other parts of the continent. From the fiels results showed that 10% of the respondents encountered trees destruction by diseases. These response figures suggest a worrying lack of preparedness in the event of more widespread disease outbreaks in near future.

4.5.8 Trees Theft

Some respondents reported that their trees were stolen in the woodlot farms near harvesting time. Due to the growing demand for wood products domestically, expansion of timber industries and carbon trade are some of potential growth factors likely to stimulate and escalate people to steal trees in their neighbours' woodlot farms. There are other reasons for trees theft, firstly the woodlot owners do not live near their woodlot farms. Another reason is that the trees theft occurs where there is an ongoing land boundary dispute or deed overlap, or where there has been a historical difference of opinion on boundary lines (Harrison et al, 2001).

4.6 Levels of Woodlot Ownership

Woodlots were also regarded as the best means of raising trees for providing logs, timbers and transmission poles to urban and rural communities in Tanzania. The Table 10 below shows the levels of woodlot ownership in Mufindi district.

Table 10 Levels of Woodlot Ownership

Level of ownership	Frequency	Percentage
Individual	67	60.9
Community	17	15.5
Household	26	23.6

4.6.1 Individual Woodlot Ownership

This is the type of ownership in which the individual people acquired land and invested in woodlot farming. The data revealed that the majority of woodlot owners in Mufindi district are individuals (67%) and they reported that it was easier for them to manage their activities individually.

4.6.2 Household Woodlot Ownership

This is the level of ownership in which a woodlot is a property of household members. The data revealed about (26%) woodlot farms in Mufindi fell in this level as it observed that it was very difficult to distribute the income equally to all household members.

4.6.3 Community Woodlot Ownership

This is the level of ownership in which a woodlot farm's management is the hands of a group or an association legally registered as a separate entity apart from its shareholders. The level of woodlot ownership has its unique administration legally registered. The data revealed that about 17% of woodlot farms in Mufindi district fell under level of ownership.

The data in Table 10 above revealed that the predominant level of woodlot ownership is individual level ownership by 67% only and only 17% of respondents organized woodlot farming in community/association basis which could help them in collective bargaining for loans from microfinance institutions in the area of study to foster their woodlot farming activities.

5. Conclusion and Policy Recommendations

Basing on the findings, the study concluded that socio-economic factors that influence woodlot farming positively and significantly are income generation expectation, land size, and assets acquisitions meanwhile education level influence woodlot farming adoption negatively and significantly. Based on the findings, this study recommends the following; *first,* the government should provide proper education to woodlot owners on trees planting (species, weeding, seeds, seedling and spacing) and fire control (establishment of fire lines) in woodlots to protect the destruction during wild fire outbreaks. The woodlot farming should be carried out professionally as it contributes to the district revenues through the issuance of forest produce harvesting licenses and transit pass from one area to another. *Second,* the rural transport infrastructures especially roads should be improved and maintained timely to enable easy transportation of goods to the market. *Third,* the government with other development partners should develop and provide capital, technical and technological support to woodlot owners in the district to accelerate the national poverty alleviation efforts which can be achieved through woodlots investment. *Fourth,* the government should look upon solving the problems facing woodlot owners like regulating prices for timbers, logs and transmission poles to enable woodlot owners to earn more income to improve their rural life. *Sixth,* proper harvesting methods, equipment as well as modern sawmilling technology should be used so as to reduce trees waste. *Seventh*, there is a need for government to support the woodlot owners by providing subsidies including inputs such as land, fertilizers, and quality seeds through Mufindi district officials so as to create employment opportunities among majority of rural households and raise their standard of living. If all these recommendations are implemented, then improvement in the income from woodlot farming would ultimately be realized.

References

Akida, A., & Blomley, R. (2006). Trend in ownership, forest resource tenure, and institution arrangement: Are they contributing to better forest management and poverty reduction. A Case Study of the United Republic of Tanzania. Retrieved from http//www.fao.org.forestry/media/12503/1/19/html.

Arnold, J.E.M. (1984). Forestation for local community development. In: Wiersum, K.F. (Ed). Strategies and designs for afforestation, reforestation and tree planting, 48-62. Proceedings of an international symposium on the occasion of 100 years of forestry education and research in the Netherlands, Wageningen, Netherlands. 432pp.

Bailey, K. D. (1994). *Methods of social research.* The Free Press, New York.

Barnes, B. V., & Wagner, W. H. (2004). *Michigan Trees: A Guide to the Trees of the Great Lakes Region.* University of Michigan Press, Michigan.

Buyinza, M and Ntakimanyire.A (2008) Rotational Woodlot Technology in Kirobya Sub – Country, Hoima District, Uganda.Botany Research Journal1 (2) 36 – 42.

Chambers R and Conway G (1992) Sustainable rural livelihoods: practical concepts for the 21st century. IDS Discussion Paper 296,

Chikoko, G. M. (2002), *A comparative analysis of household owned woodlots and fuel sufficiency between female and male headed households.* A pilot study in rural Malawi, Africa. Thesis for the Award of Doctor of Philosophy Degree Oregon State University, USA.

Churchill, G., & Lacobucci, D. (2002). *Marketing Research Methodological Foundations* (8th ed.). Harcourt Publishing, London.

Cochet, H. (2004). Agrarian dynamics, population growth and resource management: the case of Burundi. GeoJournal, 60(2), 111-122.

Creswell, J. W. (2011). *Educational Research: Planning, Conducting, and Evaluating Quantitative and Qualitative Research.* New Jersey: Pearson Education International.

Dewees, P. A., & Saxena, N. C. (1997). Wood product markets as incentives for farmer tree growing. In J. E. M. Arnold, & P. Dewees (Eds.), *Farms, trees and farmers: responses to agricultural intensification.* Earthscan, London.

Deressa TT, Hassan RM, Ringler C et al (2009) Determinants of farmers' choice of adaptation methods to climate change in the Nile Basin of Ethiopia. Global Environ Change 19(2):248–255

Dilley M. and Boudreau T.E. (2001) Coming to terms with vulnerability: a critique of the food security definition. Food Policy, Volume 26, Number 3, June 2001, pp. 229-247(19).

Edwin, S. (1990). Communal Woodlots in Tanzania-Farmers Responses and Evolving Extension Strategy-Woodlot studies in Africa. *Canadian Journal of African Studies, 2*, 282. Retrieved from http://www.odi.Org.uk/Networks/rdfn/htm.

Essa M. (2004), Household income and natural forest conservation by agroforestry: an analysis based on two agro-ecological zones: Bagrot and Jalalabad in Northern Pakistan, MSc Thesis Norwegian University of Life Sciences, Norway.

Food and Agriculture Organization (2001)

Forest Development Trust Survey (FDT). (2014). *Dar es Salaam, Tanzania.*

Foy, T. J., & Pitcher, M. J. (1999). Private sector/ community partnership for commercial Forestry development recent experience from South Africa. *The International* F*orestry Review, 4*, 222-223.

Guariguata MR, Balvanera P (2009) Tropical forest service flows: improving our understanding of the biophysical dimension of ecosystem services. For Ecol Manag 258:1825–1829.

Gujarati, D. N. (2006). *Essentials of Econometrics* (3rd ed.). McGraw-Hill.

Hair, J. F., Black, W. C., Babin, B. J., Anderson, R. E., & Tatham, R. L. (2006). *Multivariate Data Analysis*. Pearson Prentice Hall, Upper Saddle River.

Haule, E. F. (2003). *Spatial Technology for Natural Resource Management, Environmental Monitoring, and Disaster Management.* Proceeding of Regional on use of Geo-spatial technology, 2004, Khartoum, Sudan.

Harrison, S.R., Herbohn, J.L. and Emtage, N.F. (2001). Estimating investment risk in small – scale plantations of rainforest cabinet species and eucalyptus. Cheltenham, UK.

Hobbs, J. E., & Young, L. M. (2001). *Closer vertical co-ordination in agri-food supply chains: A conceptual framework and some preliminary evidence.*

https://doi.org/10.1108/13598540010338884

Kajembe, G. C., & Luoga, E. J. (1996). *Socio – economic aspects of tree farming in Njombe district.* Consultant report to HIMA – Njombe. SUA, Morogoro. pp. 43-48.

Katani, G. C. (1994). *The role of Gender based Knowledge in Developing Coping Strategies Against Deforestation: A case study of Mwanza district dissertation for the award of Msc. degree of Sokoine university of agriculture, Morogoro, Tanzania.* pp. 110.

Kimaro A.A, Timmer VR, Mugasha A.G et al (2007) Nutrient use efficiency and biomass production of tree species for rotational woodlot systems in semi-arid Morogoro, Tanzania.

Kirsten, J., Mapila, M., Okello, J., & De, S. (2012). *Managing Agricultural commercialization for inclusive growth in Sub-Saharan Africa.* Policy research paper1, The Global Development Network.

Lyne, G. D. (2002). Agricultural Industrialization: A Metaeconomics Look at the Metaphors by which we live. *Review of Agricultural Economics, Agricultural and Applied Economics Association, 24*(2), 410-427. https://doi.org/10.1111/1467-9353.00106

Makauki, A. F. (1999). *Factors Affecting Adoption of Agro forestry Farming System in Turiani Division, Morogoro Rural District.* Dissertation for an Award of M. A. Degree at Sokoine University of Agriculture, Morogoro, Tanzania. pp. 92.

Mnyawi, S. P., & Benedict, S. (2014). Contribution of Petty Trade on Household Poverty Reduction: A Case of Madukani, Dodoma Region. *Journal of Innovation and Scientific Research, 6*(2), 111-119.

Mwaipopo, L. C. (2006). *Small scale dairy farming for livelihoods of rural farmers: Constraints and Livestock Research for Rural Development.*

Muyanga, M., & Jayne, T. S. (2014). Effects of rising rural population density on smallholder agriculture in Kenya. Food Policy, 48(0), 98-113.

Nair, P.K., B.M. Kumar and V.D. Nair, (2009). Agroforestry as a strategy for carbon sequestration.J.Plant Nut.Soil Sci., 172:10 – 2.

National Bureau of Statistics Tanzania. (2013).

Ngaga, Y.M. (2011). A Platform for Stakeholders in African Forestry: Forest Plantations and Woodlots in Tanzania.

Ngailo, J. A., Mmari, W. N., & Kyando, L. S. (2011). *Soil Appraisal results from selected irrigated farmers' fields in Iringa Rural District.* Iringa District Council (IDC). Consultants' Report. pp. 34.

Nyadzi, G. I., Otsyina, R. M., Banzi, F. M., Bakengesa, S. S., Gama, B. M., Mbwambo, L., & Asenga, D. (2003). Rotational woodlot technology in northwestern Tanzania: tree species and crop performance. *Agroforestry Systems, 59*, 253-263. https://doi.org/10.1023/B:AGFO.0000005226.62766.05

Okia, C.A.,J.G. Agea, J.Sekatuba,G.Ongodia ,B.Katumba,V.I. Opolot and H.Mutabazi(2009).Candidate agro forestry technologies and practices for Uganda.Agric.J.,4:208 – 2015.

Pindyck, R. S., & Rubinfeld, D. L. (2010). *Microeconomics.* The Pearson Prentice Hall, United States.

Reed, D. (1992). *Structural Adjustment and the Environment.* London: Earthscan Publications.

Singunda, W. L. (2010). *Economic Contribution of Private Woodlots to the Economy of Mufindi District-Tanzania.* A dissertation for award of Master of Science in Forestry of Sokoine University of Agriculture. Morogoro-Tanzania. pp. 142.

Ugwumba, C. O. A., & Chukwuji, C O. (2010). The economics of catfish production in Anambra State, Nigeria.A profit function approach. *J. Agric. Soc. Sc., 6*, 105-109.

United Nations (1975). Report of the World Food Conference, Rome 5-16 November 1974. New York).

URT, (2002). Forest Act, 2002 Act No. 14 of 2002. Ministry of Natural Resources and Tourism. Dar - es- Salaam. 1281 pp.

URT (2006). Ministry of Natural Resources and Tourism Report.

Williamson, O. E. (2000). The Institutional Economics: Taking Stock, Looking Ahead. *The Journal of Economic Literature, 38*(3), 595-613. https://doi.org/10.1257/jel.38.3.595.

World Food Summit (1996).

Zolho, R. (2005). Effect of Fire Frequency on the regeneration of Miombo Woodland in Nhambita, Mozambique A dissertation presented for the degree of Master of Science University of Edinburgh, 63pp.

Notes

Note 1. Validity is a precondition of reliability (Mwaipopo, 2006), but reliability is not a precondition of validity. This argument implies that reliability is a necessary condition of validity, but it is not sufficient to establish validity (Gujarati, 2006; Pindyck and Rubinfeld, 2010).

Copyrights

Publisher: Eliva Press Global Ltd

Email: info@elivapress.com

Made in United States
Orlando, FL
04 November 2026

90200622R10024